U.S. Department
of Transportation

**National Highway
Traffic Safety
Administration**

DOT HS 809 580   **December 2004**

Technical Report

Cell Phone Use on the Roads
in 2002

Published By:
NCSA
National Center for Statistics and Analysis

1. Report No. DOT HS 809 580	2. Government Accession No.	3. Recipient's Catalog No.
4. Title and Subtitle Cell Phone Use on the Roads in 2002		5. Report Date December 2004
		6. Performing Organization Code NPO-101
7. Author(s) Glassbrenner, Donna, Ph.D.		8. Performing Organization Report No.
9. Performing Organization Name and Address Mathematical Analysis Division, National Center for Statistics and Analysis National Highway Traffic Safety Administration U.S. Department of Transportation NPO-101, 400 Seventh Street, S.W. Washington, D.C. 20590		10. Work Unit No. (TRAIS)
		11. Contract or Grant No. DTNH22-00-07001
12. Sponsoring Agency Name and Address Mathematical Analysis Division, National Center for Statistics and Analysis National Highway Traffic Safety Administration U.S. Department of Transportation NPO-101, 400 Seventh Street, S.W. Washington, D.C. 20590		13. Type of Report and Period Covered NHTSA Technical Report
		14. Sponsoring Agency Code

15. Supplementary Notes

Data collection, estimation, and variance estimation for NHTSA's National Occupant Protection Use Survey were conducted by Westat, Inc under the direction of NHTSA's National Center for Statistics and Analysis under federal contract number DTNH22-00-07001. Also we thank Mike Goodman and Julie Barker in NHTSA's Office of Advanced Safety Research, and Paul Tremont in the Office of Research and Technology for helpful comments.

Abstract

With the increasing popularity of cellular phones and public concern about the safety of using phones while driving, there has been increased interest in tracking the incidence of driver cell phone use. This report presents the National Highway Traffic Safety Administration's (NHTSA) most recent results on this topic, which come from NHTSA's National Occupant Protection Use Survey (NOPUS). The survey estimated that during daylight hours, drivers of cars, trucks, vans, and sport utility vehicles used hand-held phones during 4% of their driving time in 2002, up from 3% in 2000. These results were obtained by observing actual traffic. In fact the NOPUS is currently the only source of probability-based observed national data on driver cell phone use.

While the NOPUS percentages are small, they indicate a substantial and growing phenomenon, translating into approximately 600,000 drivers on the road at any given time between 8 AM and 6 PM using hand-held phones in 2002, up from 400,000 in 2000. When combined with data from other surveys on hands-free use, NOPUS that nearly finds that 900,000 drivers on the road at any given daylight time are using cellular phones in some manner, whether by holding the phone or using a hands-free device.

However, while NOPUS finds an increase in hand-held use behind the wheel, the survey data indicate a shift toward the use of hands-free devices. The growth in hands-free use while driving that occurred between 2000 and 2002 outpaced the growth in hand-held phones (a 33% growth for hand-held phones, compared to 100% for hands-free).

NOPUS first observed cell phone use in 2000, and so the data in this report represent some of the first direct measurements of this form of driver distraction on the road.

17. Key Words cellular phones, wireless phones, driver distraction, observational surveys, hand-held phones, hands-free devices	18. Distribution Statement Document is available to the public through the National Technical Information Service, Springfield, VA 22161 http//:www.ntis.gov		
19. Security Classif. (of this report) Unclassified	20. Security Classif. (of this page) Unclassified	21. No. of Pages 46	22. Price

Form DOT F 1700.7 (8-72)   Reproduction of completed page authorized

TABLE of CONTENTS

NCSA   National Center for Statistics and Analysis, 400 Seventh St., S.W., Washington, DC 20590

NCSA National Center for Statistics and Analysis, 400 Seventh St., S.W., Washington, DC 20590

Tables and Figures

NCSA National Center for Statistics and Analysis, 400 Seventh St., S.W., Washington, DC 20590

1. Executive Summary

With the increasing popularity of cell phones and public concern about the safety of using phones while driving, there has been increased interest in tracking the incidence of driver cell phone use. This report presents the National Highway Traffic Safety Administration's (NHTSA) most recent results on this topic, which come from NHTSA's National Occupant Protection Use Survey (NOPUS). The survey estimated that during daylight hours, drivers of cars, trucks, vans, and sport utility vehicles used hand-held phones during 4% of their driving time in 2002, up from 3% in 2000. These results were obtained by observing use in actual traffic. In fact the NOPUS is currently the only source of probability-based observed national data on driver cell phone use.

While the NOPUS percentages are small, they indicate a substantial and growing phenomenon, translating into approximately 600,000 drivers on the road at any given time between 8 AM and 6 PM using hand-held phones in 2002, up from 400,000 in 2000. When combined with data from other surveys on hands-free use, NOPUS that nearly finds that 900,000 drivers on the road at any given daylight time are using cellular phones in some manner, whether by holding the phone or using a hands-free device. (See Section 6 for the derivation of these statistics.)

However, while NOPUS finds an increase in hand-held use behind the wheel, the data also indicate a shift toward the use of hands-free devices while driving. The growth in hands-free use while driving that occurred between 2000 and 2002 outpaced the growth in hand-held phones (a 33% growth for hand-held phones, compared to 100% for hands-free). (Again see Section 6 for derivations.)

NOPUS found increases in hand-held use among drivers in the 16-24 and 25-69 age ranges and on urban roads, and continued to find similar use by males and females. NOPUS first observed cell phone use in 2000, and so the data in this report represent some of the first direct measurements of this form of driver distraction on the road.

1.1 More Drivers Are Holding Phones

Figure 1: Percent of Drivers Holding Phones

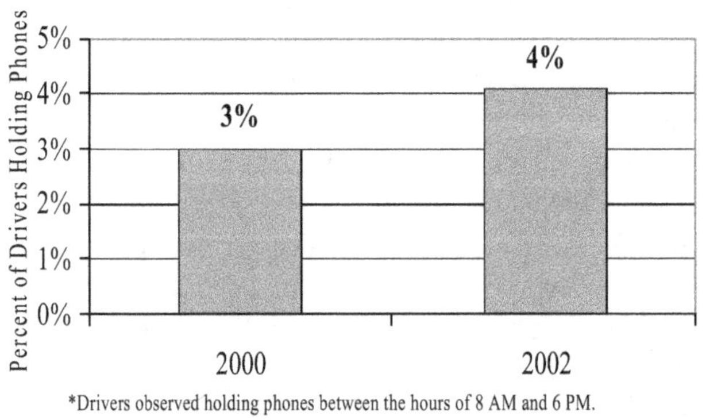

*Drivers observed holding phones between the hours of 8 AM and 6 PM.

Source: National Occupant Protection Use Survey, National Center for Statistics and Analysis. NHTSA

NCSA National Center for Statistics and Analysis, 400 Seventh St., S.W., Washington, DC 20590

In 2002, 4% of drivers observed at intersections controlled by stop signs or stoplights during daylight hours were observed holding phones, up from 3% in 2000. These results are from the National Occupant Protection Use Survey (NOPUS), which is conducted by the National Center for Statistics and Analysis in the National Highway Traffic Safety Administration (NHTSA) and provides the only probability-based observed data on driver cell phone use in the United States.

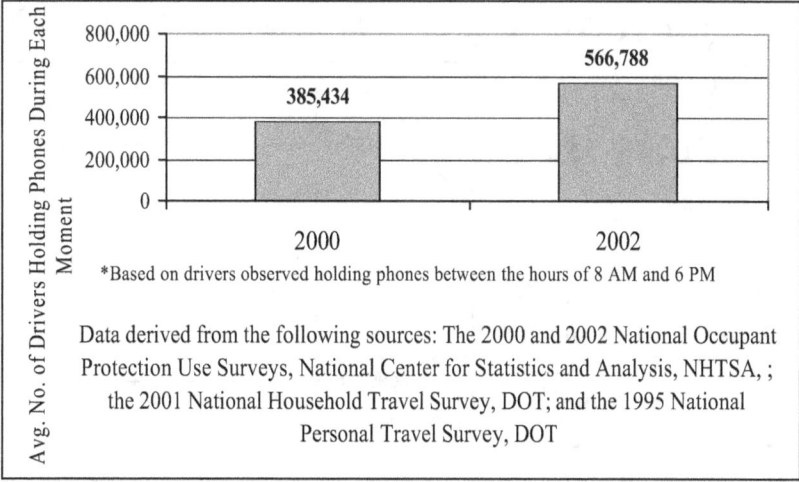

Figure 2: The Number of Drivers Holding Phones At Any Given Time

*Based on drivers observed holding phones between the hours of 8 AM and 6 PM

Data derived from the following sources: The 2000 and 2002 National Occupant Protection Use Surveys, National Center for Statistics and Analysis, NHTSA, ; the 2001 National Household Travel Survey, DOT; and the 1995 National Personal Travel Survey, DOT

While the increasing prevalence of drivers holding phones is apparent to many who drive on the nation's roads, what NOPUS offers is a scientifically-based quantification of the prevalence of this phenomenon and how much it has increased. It may surprise many that the NOPUS percentages are so low (4% in 2002), compared to their own personal observations, perhaps because drivers holding phones stand out in their minds when they observe them. (Keep in mind that the above estimates concern only hand-held phones. Hands-free estimates will follow below.) Also, we will see next that NOPUS's small estimates translate into large numbers of drivers using phones.

1.2 Over Half a Million Drivers Are Holding Phones At Any Given Time

Figure 3: Percent of Drivers Using Phones, by Means of Use

While the use rates found by NOPUS are small numbers, they indicate a sizable phenomenon because of the sheer volume of vehicles on the road. At any given time between the hours of 8 AM and 6 PM in 2002, there were about 14 million privately owned vehicles on the road. The

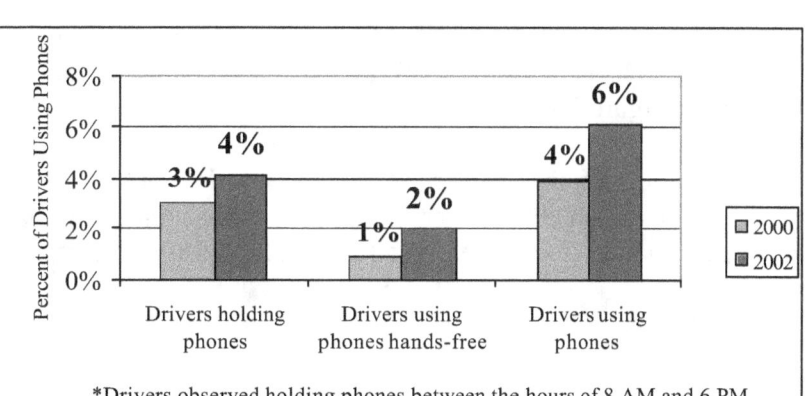

*Drivers observed holding phones between the hours of 8 AM and 6 PM.

Data derived from the following sources: The 2000 and 2002 National Occupant Protection Use Surveys, National Center for Statistics and Analysis, NHTSA; the 2001 National Household Travel Survey, DOT; the 1995 National Personal Travel Survey, DOT; and the CTIA's Semi-Annual WirelessIndustry Indices , April 2003 Edition

NCSA National Center for

NOPUS use rate of 4% means that nearly 600,000 of these vehicles were driven by someone holding a phone. (See Section 6 for the derivation of these statistics.)

1.3 The Shift Toward Hands-Free Use

Although the 2002 survey only observed <u>hand-held</u> phones, by combining its estimates with other data sources, we can estimate the extent to which drivers used phones hands-free. Using data from interviews of cell phone users, we find dramatic growth in hands-free use while driving – a 100% growth in the <u>percent</u> of drivers using phones hands-free, over the level seen in 2002, and a nearly a 150% growth in the <u>number</u> of drivers on the road at any given

Figure 4: Number of Drivers on Phones At Any Given Time, By Means of Use

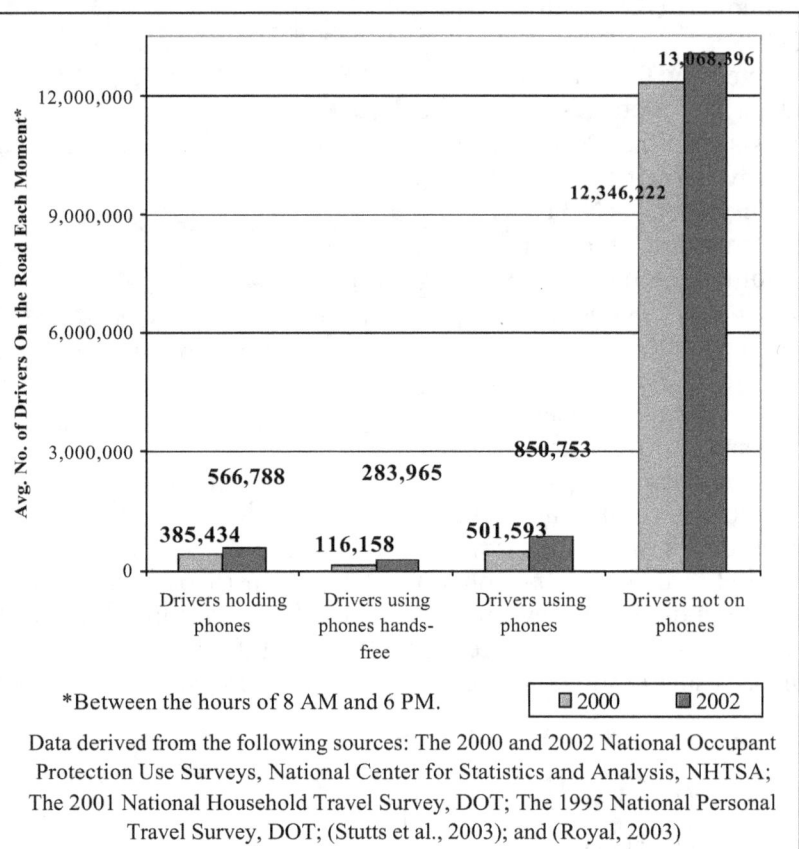

*Between the hours of 8 AM and 6 PM.

Data derived from the following sources: The 2000 and 2002 National Occupant Protection Use Surveys, National Center for Statistics and Analysis, NHTSA; The 2001 National Household Travel Survey, DOT; The 1995 National Personal Travel Survey, DOT; (Stutts et al., 2003); and (Royal, 2003)

time who are using phones in the hands-free mode. (See Figures 3 and 4. Estimates in Figure 4 do not necessarily sum to totals due to rounding.)

Note that during the same period of time, the percent of drivers on hand-held phones only increased by 33%, indicating a shift towards using hands-free phones when people are behind the wheel. (See Figure 3.)

Nearly 900,000 vehicles on the road at any given daylight time are being driven by someone on a phone, with about 600,000 drivers holding the phone and the remaining 300,000 using phones in some type of hands-free mode. (Again see Section 6 for derivations.)

National Center for Statistics and Analysis, 400 Seventh St., S.W., Washington, DC 20590

1.4 Other Major Findings of the 2002 Survey

Other major findings of the 2002 NOPUS include the following. Each of the findings concerns the use of hand-held phones by drivers during daylight hours:

- Use increased among young adults and adults, ages 16-24 and 25-69. Use is now statistically lower among older drivers than among those under 70 years of age.

- Males and females continue to exhibit similar use, with each talking on hand-held phones for 4% of their daylight driving time in 2002.

- Use increased on urban roads, with a statistically significant increase from 2% in 2000 to 5% in 2002.

Figure 5: Hand-Held Phone Use By Select Characteristics

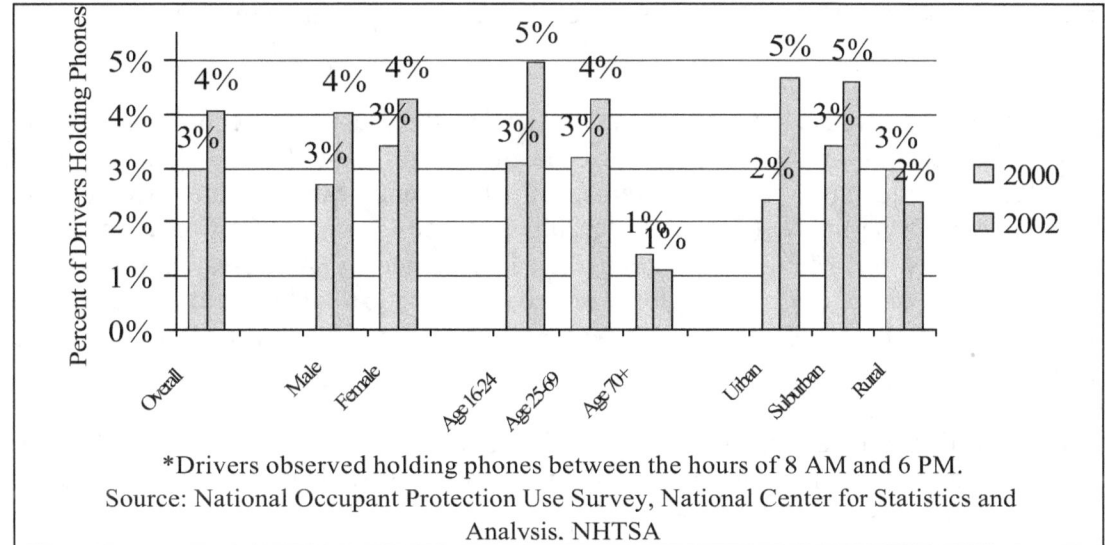

*Drivers observed holding phones between the hours of 8 AM and 6 PM.
Source: National Occupant Protection Use Survey, National Center for Statistics and Analysis. NHTSA

Report Organization

This report is organized as follows. Section 2 presents basic information concerning wireless phones that the lay reader may find helpful. Section 3 presents and analyzes the NOPUS estimates for national patterns. Section 4 examines subnational estimates, such as use by age and gender. Section 5 describes the NOPUS design and data collection methods.

The estimates that are obtained from NOPUS data in conjunction with data sources, including many estimates in this Executive Summary, will be derived in Section 6. Along the way, we will derive the independently interesting statistics in Section 6.2 that that a snapshot of drivers on phones on the road (during daytime, in 2002) would consist of an estimated 67% using hand-held phones and 33% using hands-free phone devices.

NCSA National Center for Statistics and Analysis, 400 Seventh St., S.W., Washington, DC 20590

Terminology

In this report we will use the words "wireless phone" and "cellular phone" (or "cell phone") interchangeably. We shall also sometimes simply use the term "phone" to describe either. The term "subscriber" will be used to include prepaid subscriptions as well as subscribers to service plans. See Section 2 for a description of various types of wireless phones. We will refer to the period between 8 AM and 6 PM as "daylight" hours. Entries of tables might not sum to totals due to rounding.

NOPUS Is the Official Measure of Use

Most estimates concerning driver cell phone use come from interviews of drivers, such as those in (Boyle et al., to appear), (Royal, 2003), and (Stutts, 2003). While interviews provide crucial information on aspects such as the situations in which people will use phones while driving, the NOPUS estimates are obtained from observations of actual traffic, and so better reflect actual use on the road. In fact, NOPUS provides the only probability-based observed data on driver cell phone use nationwide. In this way NOPUS provides a unique role in the study of driver cell phone use, and, in NHTSA's view, provides the best national measure of the incidence of driver cell phone use.

NOPUS first observed cell phone use in 2000, and so the data in this report represent some of the first measurements of the actual occurrence of this form of driver distraction on the road. Wireless subscriptions have risen dramatically in recent years and are expected to increase substantially in the future. (Roche et al., 2003; Yankee Group, 2002) Consequently driver cell phone use has the potential for large increases as well. As research provides a greater understanding of the risks of this form of driver distraction and several jurisdictions consider restricting cell phone use, it will be important to continue to track actual cell phone use on the roads.

Data collection, estimation, and variance estimation for the NOPUS were conducted by Westat, Inc under the direction of NHTSA's National Center for Statistics and Analysis under federal contract number DTNH22-00-07001.

NCSA National Center for Statistics and Analysis, 400 Seventh St., S.W., Washington, DC 20590

2. Background: A Primer on Wireless Phone Technology.

For the lay reader's information we include some basic information on wireless technology, at least as it existed in 2000 and 2002, the primary data years for this report.

First we note that different people might consider different technologies to be a "wireless phone", a "cell phone", both, or neither. For technical definitions of cellular phones, wireless phones, satellite phones, and other wireless communications, the reader should consult documents by the Federal Communications Commission (FCC), e.g. on their website www.fcc.gov.

By far the most common type of wireless phone is the cellular phone – a portable phone that transmits speech and other information through a network of ground-based cell towers. Cellular phones may be held by hand or may be used with a hands-free attachment such as an earpiece with an attached microphone or a headset. In addition, some cell phones come with a "push-to-talk" feature, in which the user can use the phone e.g. in his/her lap by pushing a button when s/he wishes to speak. The user can listen to the other party on the phone by releasing the button.

Another communications technology that can be used in a vehicle is a car phone, which is a phone installed and integrated into the car. Car phones either attach via cords to the dashboard, or utilize speakers and microphones in the dashboard, such as the Onstar system. In particular, car phones might be utilized in either a hand-held or hands-free mode.

A satellite phone is a hand-held phone that transmits speech and other information via satellite. These phones are generally larger than cellular phones, and may have hands-free attachments. However because they are expensive and are used mostly by the military, it would be rare to observe a driver using a satellite phone on U.S. roads.

Other wireless phones are incorporated into certain Personal Data Assistants (PDAs), such as Blackberries. These transmit information via cell phone towers, and may have hands-free attachments such as earbuds or headsets. However this technology was relatively rare during the time the 2000 and 2002 NOPUS were conducted.

What Does "Hand-Held Use" Mean?

The notion of "using a hand-held phone" might be interpreted by different people as comprising different sets technologies (e.g. is "push-to-talk" a hand-held technology?) or different sets of behaviors (does checking email on a PDA constitute "using a phone"?). Some people might consider "push-to-talk" hands-free use in the context of driving a vehicle, since one is only required to touch the phone when one is speaking. Some people might consider checking email on a PDA "using a phone" since they are using their PDA and the PDA has phone capabilities.

NCSA   National Center for Statistics and Analysis, 400 Seventh St., S.W., Washington, DC 20590

Some people might even limit their notion of "using a phone" to the activities of conversing, i.e. speaking and listening to a party on the other end of the line, and so not consider dialing to be "using a phone".

Since this report presents a variety of hand-held use estimates from different sources, it is worth summarizing what "hand-held use" means in each context. Additional detail are supplied in the methodology sections of this report.

The NOPUS Estimates

The majority of the estimates in this report on hand-held use, and indeed all of the hand-held estimates presented by this report as new, are produced by the NOPUS survey. The survey data were obtained by data collectors observing vehicles stopped in actual traffic on the roads. Data collectors classified a driver as "using a hand-held phone" if <u>the driver was holding what appeared to be a phone to his/her ear</u>. Observers were not trained in the various types of wireless phones, and so may or may not have counted technologies such as PDAs or corded car phones (or, in the event they saw one, a satellite phone) as phones, depending on their interpretation of what constitutes a "phone".

This definition of "hand-held use" was implemented in NOPUS because it is a simple observable definition that captures much hand-held activity. Note however that this definition gives rise to some odd characterizations. Activities such as dialing with one's fingers, which could be difficult from the roadside, would <u>not</u> be counted as "using a hand-held phone" in NOPUS (unless the driver were somehow manually dialing with the phone to his/her ear), while voice-activated dialing <u>would</u> constitute use.

In summary, the NOPUS hand-held estimates reflect the types of phones subjectively viewed by the data collectors as "phones", and drivers are characterized as "using a hand-held phone" if they are holding the "phone" to their ear. Additional detail are provided in Section 5.

Other Data Sources

This report also cites estimates on hand-held use from other data sources, namely from (Boyle et al., 2001), (Royal, 2003), and (Stutts et al., 2003). These data are used for the sole purpose of constructing hands-free estimates from the NOPUS hand-held estimates.

These additional surveys were conducted via telephone interviews of respondents, and so relied to some extent on the respondents' notions of what constitutes "using a hand-held phone". In some cases, survey questions might have mentioned particular types of phones, telecommunications service (such as GSM), or communications protocols (such as PCS), and this information might have influenced the respondents' answers. The reader is encouraged to consult these surveys for the specific wording of questions.

In particular, the estimates of hand-held use from these surveys might reflect various types of wireless phones, depending on the survey questions and respondents' notions of hand-held use.

NCSA National Center for Statistics and Analysis, 400 Seventh St., S.W., Washington, DC 20590

What Constitutes "Hands-Free Use"?

As with hand-held use, the notion of "using a phone hands-free" can mean different things to different people. Is "push-to-talk" hands-free? Is dialing considered "use"?

The estimates in this report concerning hands-free phone use derive from the NOPUS estimates of hand-held use and telephone surveys (Boyle et al., 2001), (Royal, 2003), and (Stutts et al., 2003). Again the telephone surveys necessarily relied in part on respondents' notions of wireless phones, what they consider use, and what use behaviors they consider to be hands-free. The reader should consult the individual surveys for additional detail.

NCSA National Center for Statistics and Analysis, 400 Seventh St., S.W., Washington, DC 20590

3. Driver Cell Phone Use Nationwide

This section provides information that further illuminates the NOPUS nationwide use rates. Section 3.1 discusses the national estimates of hand-held use from NOPUS, and Section 3.2 presents derived estimates on the use of cell phones through any means, whether by holding the phone or using a hands-free device. Section 3.3 compares the NOPUS estimates with those from other surveys.

3.1 The NOPUS National Estimate

NOPUS found that hand-held use increased from 3% in 2000 to 4% in 2002. That is, in 2002, 4% of drivers stopped at a stop sign or stoplight between the hours of 8 AM and 6 PM were holding a phone. This estimate reflects a "snapshot" of use, but one can also infer from it that drivers were using hand-held phones for 4% of their driving time in 2002.

Table 1: Cell Phone Use Nationwide

The Percent of Drivers Holding Cell Phones[1]	In 2002		In 2000		2000-2002 Change	
	Estimate	Standard Error	Estimate	Standard Error	Estimate	Standard Error
	4%	0.7%	3%	0.5%	1%	0.8%

[1] Drivers of passenger vehicles with no commercial markings observed between 8 AM and 6 PM at intersections controlled by a stop sign or stoplight.
Source: National Occupant Protection Use Survey, National Center for Statistics and Analysis, NHTSA

The increase in hand-held use, from 3% in 2000 to 4% in 2002 is statistically significant with 80% confidence. That is, we are fairly confident (80% confident) that use increased. In addition the measured increases that NOPUS saw in virtually all of the subcategories in which it observed use, such as vehicle type, age of drivers, and urbanization, are also consistent with greater use nationwide. (See Section 4 for the NOPUS subnational estimates.)

We note that during the period 2000-2002, one state (New York)

Figure 6: Cell Phone Use Nationwide

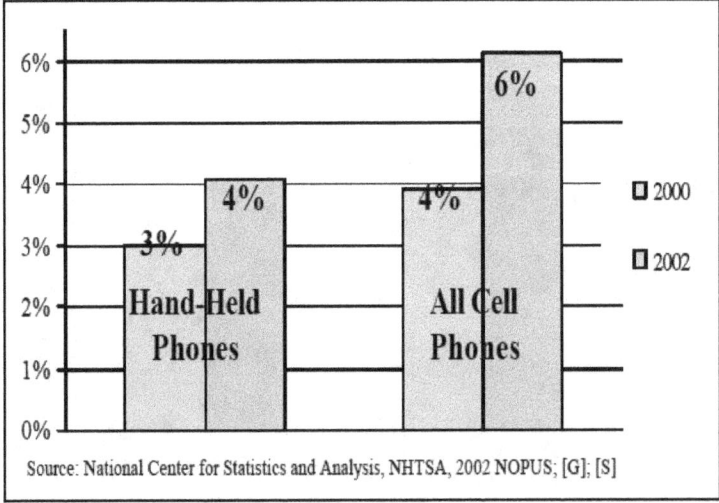

Source: National Center for Statistics and Analysis, NHTSA, 2002 NOPUS; [G]; [S]

passed a law banning the use of hand-held phones while driving. It is possible that the nationwide use of hand-held phones would have increased more without the NY ban. If more jurisdictions restrict driver cell phone use, this could impact future use rates.

In fact when the NOPUS observations were made in June 2002, the only state that substantially restricted driver cell phone use was New York, which in November 2001 banned the use of hand-held phones while driving. Other jurisdictions such as New Jersey and the District of Columbia have since enacted hand-held bans (New Jersey in May 2004 and D.C. in July 2004), but these newer laws would not have affected the NOPUS data, which was last collected in 2002. A few other states, such as Massachusetts, allow cell phone use as long as it does not interfere with the operation of the vehicle. Some major cities, such as Santa Fe, have hand-held bans similar to that in New York. (Governor's Highway Safety Association website) Many jurisdictions, however, have no or relatively weak restrictions on cell phone use, and this might explain to some extent why use is as prevalent as it is.

3.2 Estimates of the Use of All Wireless Phones

Using auxiliary data sources, one can extrapolate the NOPUS estimates, which only concern hand-held phones, to all wireless phones. Doing this, we estimate that 6% of drivers stopped at a stop sign or stoplight between the hours of 8 AM and 6 PM in 2002 were using some type of wireless phone, compared to 4% in 2000. These estimates are derived in Section 6.

3.3 Comparing the NOPUS Findings to Results of Other Surveys

Although NOPUS's estimates might initially seem low, they actually indicate use to be quite substantial. NOPUS found that drivers are using cell phones on average 6% of the time during daylight hours. The percentage of trips in which a driver used a phone, and the percentage of drivers who use phones during some substantial fraction of their trips are bound to be higher, assuming most phone conversations are short compared to the duration of the trip.

Indeed NOPUS's results are consistent with higher use rates measuring different aspects of use in other surveys. A survey conducted by the Gallup Organization in 2002 for NHTSA found that drivers who own cell phones made outgoing calls on 18% of their trips on average, and took incoming calls on 17% of their trips. (Royal, 2003) Since

Figure 7: The Growth in Wireless Subscriptions

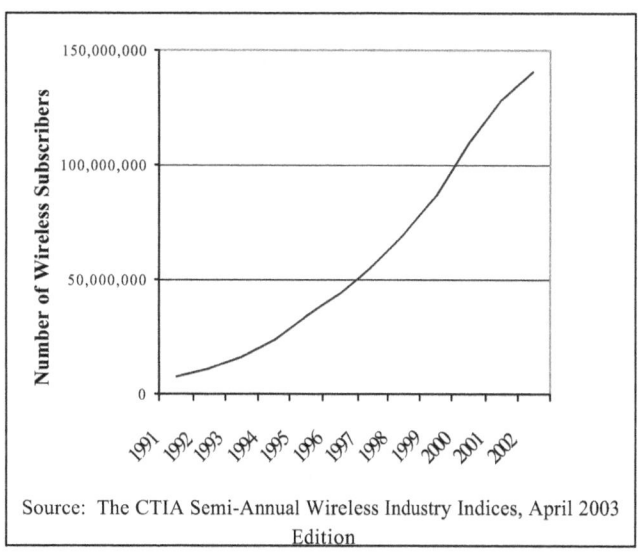

Source: The CTIA Semi-Annual Wireless Industry Indices, April 2003 Edition

NCSA National Center for Statistics and Analysis, 400 Seventh St., S.W., Washington, DC 20590

many trips during which a driver makes or receives a call also involve periods of time in which no call is being made, it is quite conceivable that calls were made 6% of the <u>time</u> that drivers were on the road. In addition, since the data in (Royal, 2003) were obtained from telephone interviews, rather than from observation, the estimates fom (Royal, 2003) are susceptible to errors in recollection. It is possible that respondents perceive their use to be greater or less than it is. The survey in (Royal, 2003) also found that 13% of <u>drivers</u> made outgoing calls on at least ¾ of their trips, with 13% taking incoming calls. The average length of a call, including both incoming and outgoing calls, was found to be 4.5 minutes.

Figure 8: Growth in the Use of Wireless Devices

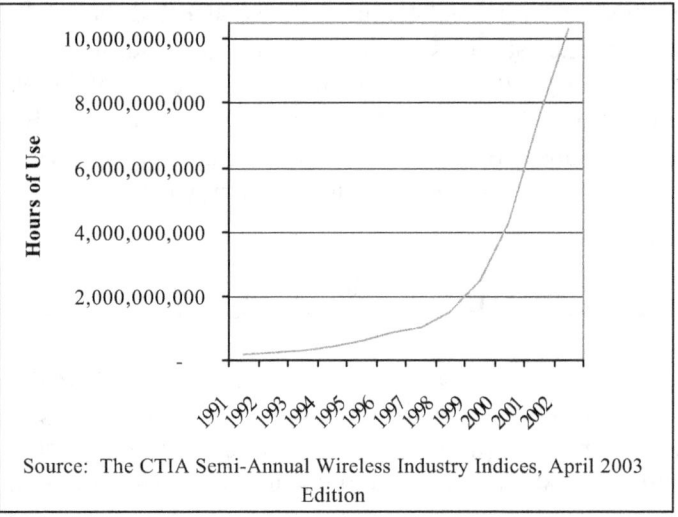

Source: The CTIA Semi-Annual Wireless Industry Indices, April 2003 Edition

The NOPUS findings also mirror growth in wireless use in general, as reported by the wireless industry. The Cellular Telecommunications and Internet Association reports dramatic increases in the number of wireless subscribers and in the amount of time that wireless devices are used. (We note that wireless devices include devices such as pagers and PDAs, as well as cellular phones.) Wireless subscriptions increased by 40% from 2000 to 2002, from 97 million in June 2000 to 135 million subscriptions in June 2002. The total time that wireless devices were used also increased dramatically during the same time period, from 4 billion hours in 2000 to 10 billion in 2002. (Source: The CTIA's Semi-Annual Wireless Industry Indices, April 2003 Edition)

NCSA National Center for Statistics and Analysis, 400 Seventh St., S.W., Washington, DC 20590

4. Subnational Estimates

This section presents use rates for hand-held phones, broken out in various ways, such as by gender, age, and geographic region. Among these, we only have data on the hand-held / hands-free distribution for gender. Since the distribution in a subcategory might differ substantially from the ratios used to derive the 2000 and 2002 national estimates of the use of hands-free cell phone devices in Section 6, we only derive subcategory estimates of hands-free use for gender. For instance, we estimate in Table 17 that drivers in the West <u>hold</u> phones during 5% of their daylight driving time, but we do **not** estimate how often cell phones are used hands-free in the West.

4.1 Changes in Age Patterns

Young drivers, ages 16-24, increased their daytime use of hand-held phones from 3% in 2000 to 5% in 2002, and drivers in the 25-69 age range increased from 3% use to 4% use. Although these increases are not statistically significant, they were large enough to make hand-held use among older drivers (ages 70 and higher), which remained constant at 1%, significantly lower than use among 16-69 year olds. See Table 6 for these estimates.

In 2000, no differences in use were detected by age, but the 2000 survey might have observed at too few sites to detect a difference. Note that age is assessed subjectively by the NOPUS data collectors, and this may affect the NOPUS age estimates to an unmeasurable degree.

4.2 Males Use Phones Hands-Free Slightly More than Women

Both male and female drivers used hand held phones for 4% of their daylight driving time in 2002, up from 3% for each group in 2000. This is fairly strong evidence that males and females engage in this behavior with the same frequency. We will see later in this section, however, that men spend more daylight time driving than women (see Table 4), and so the NOPUS results indicate that the total daylight driving time that men spend on hand-held phones substantially exceeds that spent by women. (See Table 5.)

Because we have a hand-held / hands-free distribution for men and women, we can calculate estimates of the use of all cell phones by gender. By conducting a survey that interviewed respondents, (Royal, 2003) finds hands-free use to be more popular with men.

NCSA National Center for Statistics and Analysis, 400 Seventh St., S.W., Washington, DC 20590

Table 2: The Usual Means of Cell Phone Use for Male and Female Drivers

Means of Use	Percent of Drivers Who Usually Use the Cell Phone via the Means of Use (When They Use a Phone While Driving)[1]	
	Men	Women
Holding the Phone	61%	69%
Using a Hands-Free Device	39%	31%

[1] Among drivers who use phones on at least 25% of their trips. Unknown responses have been distributed.
Source: (Royal, 2003)

In Section 6, we will use auxiliary data, namely from (Stutts et al., 2003), on the amount of time that part-time hands-free users use hands-free devices to obtain an extrapolation of the NOPUS national hand-held estimate to an estimate of hands-free use. (In this paper, we refer to people who use phones hands-free at least part of the time "part-time hands-free users".) If we had such auxiliary data broken out by gender, we could use the same technique to derive estimates of hands-free use for males and females.

Unfortunately (Stutts et al., 2003) does not break out their data by gender, so this approach was not possible. However we can obtain cruder estimates of male and female hands-free use by adjusting the NOPUS estimates by the data in the previous table. Doing this, we would estimate that approximately 3% of male drivers on the road during the daytime are using phones via hands-free devices, since 3% = 4% x 39/61. The corresponding calculation for females estimates that females drivers use hands-free devices for 2% of their driving time.

Table 3: Driver Cell Phone Use in 2002 by Gender and Means of Use

Means of Use	Driver Gender	
	Male	Female
Holding the Phone	4%	4%
Using a Hands-Free Device	3%	2%
Total	7%	6%

Data derived from the following sources:
- The 2002 National Occupant Protection Use Survey, National Center for Statistics and Analysis, NHTSA
- (Royal, 2003)

That is, hands-free units are sufficiently more popular with men to make our (crude) gender estimates of hands-free use while driving different. We do not have sufficient statistical data to determine whether this difference is statistically significant.

We do not have data on the hand-held / hands-free distribution by gender for the 2000 data year, so we cannot make the corresponding calculation from the previous table for the 2000 NOPUS data.

NCSA National Center for Statistics and Analysis, 400 Seventh St., S.W., Washington, DC 20590

We can also calculate the total daylight time spent on the phone while driving, by gender. Using the online survey calculator on the website of the National Household Travel Survey, we calculated that men drove for a total of 26 billion daylight hours during the course of 2002, compared to 21 billion hours for women.

Table 4: Daylight Driving Time in 2002, by Gender

Driver Gender	Total Time Spent Driving Between the Hours of 8 AM and 6 PM Each Day, in Millions of Hours			
	In 1995	In 2001	Average Annualized Increase	In 2002
Male	22,394	25,398	2%	25,936
Female	16,367	19,924	3%	20,588

Data derived from the following sources:
- The 2001 National Household Travel Survey, Dept. of Transportation
- The 1995 National Personal Travel Survey, Dept. of Transportation

Applying our cell phone use rates, we find that the number of hours spent by men on phones while driving in daylight exceeds that for women. The estimates that include hands-free devices in the following table are somewhat crude since they are based on a somewhat crude estimate of hands-free use by gender.

Table 5: Time Spent by Men and Women Driving While on the Phone in 2002

Daylight Driving Time Each Day, in Millions of Hours[1]	Men	Women
Total driving time[1], in millions of hours	25,936	20,588
while holding a phone[2]	1,044	879
while using a phone hands-free[2]	667	395
while using a phone[2]	1,711	1,274
while not using a phone[2]	22,514	18,040

[1] Total duration of all trips in privately owned vehicles that start between the hours of 8 AM and 6 PM.
[2] Based on observations of passenger vehicles with no commercial or government markings at intersections governed by a stop sign or stoplight.

Data derived from the following sources:
- The 2002 National Occupant Protection Use Survey, National Center for Statistics and Analysis, NHTSA
- The 2001 National Household Travel Survey, Dept. of Transportation
- The 1995 National Personal Travel Survey, Dept. of Transportation

Since men are involved in fatal crashes at a higher rate than women, their use of potentially distracting devices could result in greater numbers of fatalities and serious injuries. In 2001, 63% of occupants of vehicles in fatal crashes were male. (This statistic was calculated from NHTSA's Fatality Analysis Reporting System, which contains data on all fatal motor vehicle crashes on public roadways in the U.S.)

(Boyle et al., 2001) found that among cell phone users in 2000, men spent more time on their phones than women, talking for an average of 589 minutes per month, compared to 394 minutes for women. Consequently men spent on average 49% more time on wireless phones.

NCSA National Center for Statistics and Analysis, 400 Seventh St., S.W., Washington, DC 20590

4.3 Increased Use in Urban Areas

Driver hand-held cell phone use in urban areas increased from 2% of daylight driving time in 2000 to 5% in 2002, a statistically significant increase. (See Figure 9.) Use is now significantly lower in rural areas than in urban and suburban areas. It should be kept in mind that urbanization is assessed by the data collectors on scene, as opposed to from independent sources, such as data on population density, and so these estimates may reflect the subjectivity of the data collectors. Also note that NOPUS estimates use by drivers observed in urban areas, not the use by drivers who live in urban areas.

4.4 Increases in Pickup Trucks, in the West, and Among Minorities

Although we cannot assert an increase with 95% confidence, we are fairly confident that daytime use of hand-held phones increased in pickup trucks (93% confidence), the West (90% confidence), blacks (91% confidence), and other races (89% confidence). (See Figure 10.)

NCSA National Center for Statistics and Analysis, 400 Seventh St., S.W., Washington, DC 20590

4.5 Additional Hand-Held Estimates

Figure 9: Driver Cell Phone Use, by Urbanization, Region, and Time of Day and Week

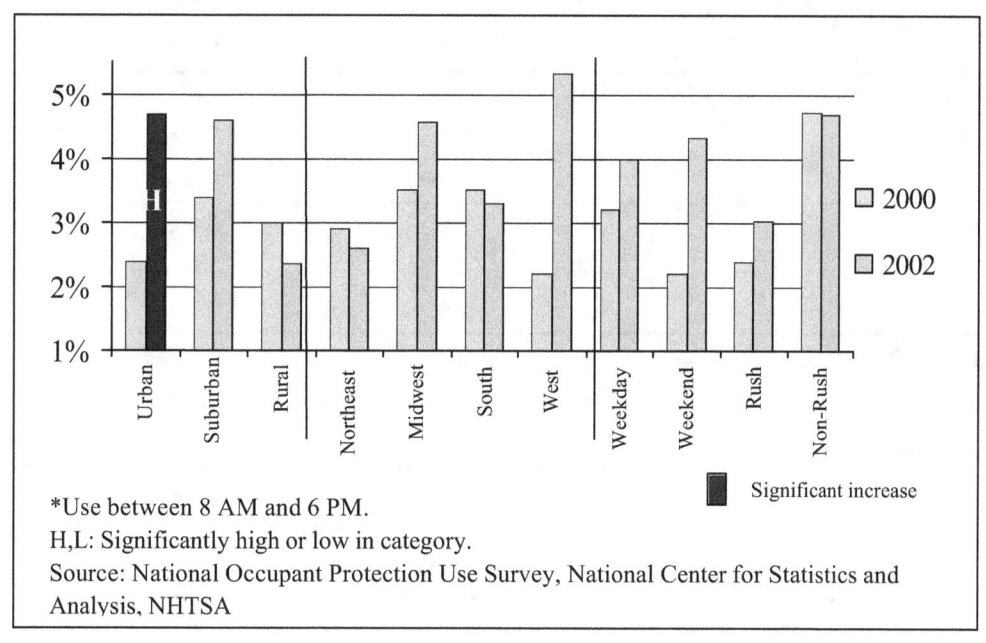

*Use between 8 AM and 6 PM.
H,L: Significantly high or low in category.
Source: National Occupant Protection Use Survey, National Center for Statistics and Analysis, NHTSA

Figure 10: Hand-Held Use, by Gender, Race, and Vehicle Type

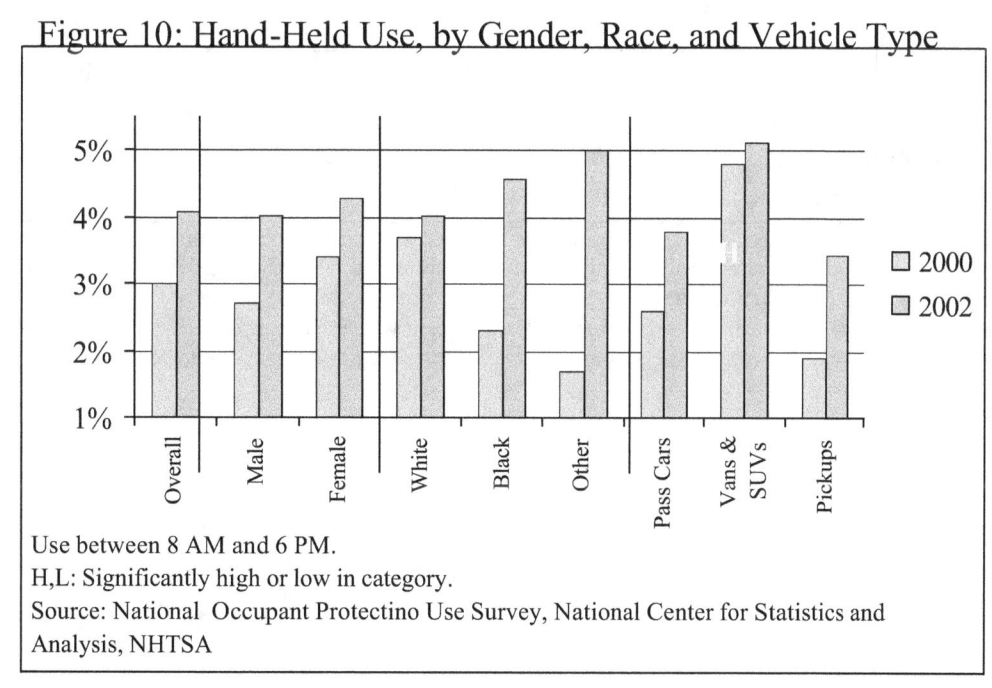

Use between 8 AM and 6 PM.
H,L: Significantly high or low in category.
Source: National Occupant Protectino Use Survey, National Center for Statistics and Analysis, NHTSA

NCSA National Center for Statistics and Analysis, 400 Seventh St., S.W., Washington, DC 20590

Table 6: Driver Hand-Held Cell Phone Use During Daylight Hours by Select Characteristics

Characteristic	Use in 2002		Use in 2000		2000-2002 Change	
	Estimate	Standard Error	Estimate	Standard Error	Estimate[2]	Standard Error
Overall	4%	0.7%	3%	0.5%	1%	0.8%
Male	4%	0.7%	3%	0.5%	1%	0.9%
Female	4%	0.8%	3%	0.6%	1%	1.0%
Age 16-24	5%	1.0%	3%	0.8%	2%	1.2%
Age 25-69	4%	0.8%	3%	0.5%	1%	0.9%
Age 70+	1% (L)	0.4%	1%	0.4%	0%	0.5%
White	4%	0.6%	4%	0.8%	0%	1.0%
Black	5%	1.0%	2%	0.9%	3%	1.3%
Other	5%	2.0%	2%	0.7%	3%	2.1%
Passenger Cars	4%	0.7%	3%	0.5%	1%	0.9%
Vans & SUVs	5%	0.7%	5%	1.0%	0%	1.2%
Pickup Trucks	3%	0.6%	2%	0.6%	1%	0.9%
Urban	5%	0.7%	2%	0.7%	3% (S)	1.0%
Suburban	5%	0.9%	3%	0.8%	2%	1.2%
Rural	2% (L)	0.6%	3%	0.8%	-1%	1.0%
Northeast	3%	0.5%	3%	1.2%	0%	1.3%
Midwest	5%	1.7%	4%	1.1%	1%	2.0%
South	3%	0.5%	4%	0.9%	-1%	1.0%
West	5%	1.8%	2%	0.7%	3%	1.9%
Weekday	4%	0.7%	3%	0.6%	1%	0.9%
Weekend	4%	1.5%	2%	0.4%	2%	1.5%
Weekday Rush Hour	3%	0.4%	2%	0.5%	1%	0.6%
Weekday Non-Rush Hour	5%	0.9%	5%	0.9%	0%	1.3%

[1] Estimated cell phone use among drivers of passenger vehicles with no commercial markings observed between 8 AM and 6 PM at intersections controlled by a stop sign or stoplight. Estimates marked with an "H" or "L" are significantly high or low in their category.

[2] Change estimates marked with an "S" are statistically significant with 95% confidence.

Source: National Occupant Protection Use Survey, National Center for Statistics and Analysis, NHTSA

NCSA   National Center for Statistics and Analysis, 400 Seventh St., S.W., Washington, DC 20590

Table 7: Driver Hand-Held Cell Phone Use During Daylight Hours by Gender and Vehicle Type

Characteristic	Use in 2002		Use in 2000		2000-2002 Change	
	Estimate	Standard Error	Estimate		Estimate	Standard Error
Males	4%	0.7%	3%	0.5%	1%	0.9%
Passenger Cars	4%	0.9%	3%	1.0%	1%	1.4%
Vans & SUVs	4%	0.7%	3%	0.8%	1%	1.1%
Pickup Trucks	3%	0.5%	3%	0.8%	0%	0.9%
Females	4%	0.8%	3%	0.6%	1%	1.0%
Passenger Cars	4%	0.7%	3%	0.6%	1%	0.9%
Vans & SUVs	6%	0.9%	6%	1.7%	0%	1.9%
Pickup Trucks	4%	1.7%	1% (L)	0.4%	3%	1.7%

[1] Estimated cell phone use among drivers of passenger vehicles with no commercial markings observed between 8 AM and 6 PM at intersections controlled by a stop sign or stoplight. Estimates marked with an "H" or "L" are significantly high or low in their category.

[2] Change estimates marked with an "S" are statistically significant with 95% confidence.

Source: National Occupant Protection Use Survey, National Center for Statistics and Analysis, NHTSA

Table 8: Driver Hand-Held Cell Phone Use During Daylight Hours by Driver Age and Vehicle Type

Characteristic	Use in 2002		Use in 2000		2000-2002 Change	
	Estimate	Standard Error	Estimate		Estimate	Standard Error
Age 16-24	5%	1.0%	3%	0.8%	2%	1.2%
Passenger Cars	5%	1.1%	3%	0.8%	2%	1.3%
Vans & SUVs	4%	1.2%	6%	2.6%	-2%	2.9%
Pickup Trucks	5%	1.6%	1%	0.7%	4% (S)	1.7%
Age 25-69	4%	0.8%	3%	0.5%	1%	0.9%
Passenger Cars	4%	0.9%	3%	0.6%	1%	1.1%
Vans & SUVs	5%	0.6%	5%	1.2%	0%	1.4%
Pickup Trucks	4%	0.9%	3%	1.1%	1%	1.4%
Age 70+	1%	0.4%	1%	0.4%	0%	0.5%
Passenger Cars	1%	0.3%	1%	0.5%	0%	0.6%
Vans & SUVs	2%	1.1%	5%	3.2%	-3%	3.4%
Pickup Trucks	1%	0.5%	1%	0.5%	0%	0.7%

[1] Estimated cell phone use among drivers of passenger vehicles with no commercial markings observed between 8 AM and 6 PM at intersections controlled by a stop sign or stoplight. Estimates marked with an "H" or "L" are significantly high or low in their category.

[2] Change estimates marked with an "S" are statistically significant with 95% confidence.

Source: National Occupant Protection Use Survey, National Center for Statistics and Analysis, NHTSA

NCSA   National Center for Statistics and Analysis, 400 Seventh St., S.W., Washington, DC 20590

Table 9: Driver Hand-Held Cell Phone Use During Daylight Hours by Race and Vehicle Type

Characteristic	Use in 2002		Use in 2000		2000-2002 Change	
	Estimate	Standard Error	Estimate		Estimate	Standard Error
White	4%	0.6%	4%	0.8%	0%	1.0%
Passenger Cars	4%	0.7%	4%	0.9%	0%	1.1%
Vans & SUVs	5%	0.6%	5%	1.1%	0%	1.2%
Pickup Trucks	4%	0.7%	2%	0.6%	2% (S)	0.9%
Black	5%	1.0%	2%	0.9%	3%	1.3%
Passenger Cars	4%	1.0%	1%	0.7%	3% (S)	1.2%
Vans & SUVs	6%	2.0%	4%	2.2%	2%	3.0%
Pickup Trucks	5%	2.5%	1%	0.6%	4%	2.6%
Other	5%	2.0%	2%	0.7%	3%	2.1%
Passenger Cars	5%	2.6%	1%	0.7%	4%	2.7%
Vans & SUVs	6%	1.6%	3%	2.9%	3%	3.3%
Pickup Trucks	3%	0.9%	10%	6.7%	-7%	6.8%

[1] Estimated cell phone use among drivers of passenger vehicles with no commercial markings observed between 8 AM and 6 PM at intersections controlled by a stop sign or stoplight. Estimates marked with an "H" or "L" are significantly high or low in their category.

[2] Change estimates marked with an "S" are statistically significant with 95% confidence.

Source: National Occupant Protection Use Survey, National Center for Statistics and Analysis, NHTSA

Table 10: Driver Hand-Held Cell Phone Use During Daylight Hours by Urbanization and Vehicle Type

Characteristic	Use in 2002		Use in 2000		2000-2002 Change	
	Estimate	Standard Error	Estimate		Estimate	Standard Error
Urban	5%	0.7%	2%	0.7%	3% (S)	1.0%
Passenger Cars	4%	0.6%	3%	0.8%	1%	1.0%
Vans & SUVs	5%	1.2%	3%	1.3%	2%	1.8%
Pickup Trucks	6%	1.6%	2%	1.4%	4% (S)	2.1%
Suburban	5%	0.9%	3%	0.8%	2%	1.2%
Passenger Cars	4%	1.0%	3%	0.7%	1%	1.2%
Vans & SUVs	6%	1.0%	6%	1.7%	0%	2.0%
Pickup Trucks	4%	0.8%	1% (L)	0.4%	3% (S)	0.9%

NCSA National Center for Statistics and Analysis, 400 Seventh St., S.W., Washington, DC 20590

Characteristic	Use in 2002		Use in 2000		2000-2002 Change	
	Estimate	Standard Error	Estimate		Estimate	Standard Error
Rural	2%	0.6%	3%	0.8%	-1%	1.0%
Passenger Cars	2%	0.6%	2%	0.9%	0%	1.1%
Vans & SUVs	3%	0.8%	7%	2.4%	-4%	2.5%
Pickup Trucks	2%	0.6%	3%	1.5%	-1%	1.6%

[1] Estimated cell phone use among drivers of passenger vehicles with no commercial markings observed between 8 AM and 6 PM at intersections controlled by a stop sign or stoplight. Estimates marked with an "H" or "L" are significantly high or low in their category.

[2] Change estimates marked with an "S" are statistically significant with 95% confidence.

Source: National Occupant Protection Use Survey, National Center for Statistics and Analysis, NHTSA

Table 11: Driver Hand-Held Cell Phone Use During Daylight Hours by Region and Vehicle Type

Characteristic	Use in 2002		Use in 2000		2000-2002 Change	
	Estimate[1]	Standard Error	Estimate[1]		Estimate[1]	Standard Error
Northeast	3%	0.5%	3%	1.2%	0%	1.3%
Passenger Cars	2%	0.5%	3%	1.0%	-1%	1.1%
Vans & SUVs	3%	0.9%	3%	2.2%	0%	2.4%
Pickup Trucks	2%	0.3%	1%	0.6%	1% (S)	0.7%
Midwest	5%	1.7%	4%	1.1%	1%	2.0%
Passenger Cars	5%	1.7%	2%	0.8%	3%	1.9%
Vans & SUVs	5%	2.3%	6%	2.5%	-1%	3.4%
Pickup Trucks	3%	1.1%	4%	2.7%	-1%	2.9%
South	3%	0.5%	4%	0.9%	-1%	1.0%
Passenger Cars	3%	0.4%	3%	1.0%	0%	1.1%
Vans & SUVs	5%	0.8%	7%	2.1%	-2%	2.3%
Pickup Trucks	3%	0.7%	1%	0.6%	2% (S)	0.9%
West	5%	1.8%	2%	0.7%	3%	1.9%
Passenger Cars	5%	2.1%	2%	0.9%	3%	2.3%
Vans & SUVs	6%	1.3%	3%	1.4%	3%	1.9%
Pickup Trucks	4%	1.6%	2%	1.2%	2%	2.0%

[1] Estimated cell phone use among drivers of passenger vehicles with no commercial markings observed between 8 AM and 6 PM at intersections controlled by a stop sign or stoplight. Estimates marked with an "H" or "L" are significantly high or low in their category.

[2] Change estimates marked with an "S" are statistically significant with 95% confidence.

Source: National Occupant Protection Use Survey, National Center for Statistics and Analysis, NHTSA

NCSA National Center for Statistics and Analysis, 400 Seventh St., S.W., Washington, DC 20590

Table 12: Driver Hand-Held Cell Phone Use During Daylight Hours by Time of Day/Week and Vehicle Type

Characteristic	Use in 2002		Use in 2000		2000-2002 Change	
	Estimate[1]	Standard Error	Estimate[1]		Estimate[1]	Standard Error
Weekday	4%	0.7%	3%	0.6%	1%	0.9%
Passenger Cars	4%	0.8%	3%	0.6%	1%	1.0%
Vans & SUVs	5%	0.7%	6%	1.5%	-1%	1.6%
Pickup Trucks	4%	0.7%	2%	0.6%	2% (S)	0.9%
Weekend	4%	1.5%	2%	0.4%	2%	1.5%
Passenger Cars	4%	1.7%	2%	0.8%	2%	1.9%
Vans & SUVs	6%	2.0%	1%	0.7%	5% (S)	2.1%
Pickup Trucks	2%	0.7%	3%	1.9%	-1%	2.0%
Weekday Rush Hour	3%	0.4%	2%	0.5%	1%	0.6%
Passenger Cars	3%	0.4%	2%	0.5%	1%	0.7%
Vans & SUVs	4%	0.6%	5%	1.5%	-1%	1.6%
Pickup Trucks	3%	0.5%	1%	0.5%	2% (S)	0.7%
Weekday Non-Rush Hour	5%	0.9%	5%	0.9%	0%	1.3%
Passenger Cars	4%	1.0%	4%	0.9%	0%	1.4%
Vans & SUVs	6%	1.0%	8%	2.5%	-2%	2.7%
Pickup Trucks	3%	0.8%	3%	1.6%	0%	1.8%

[1] Estimated cell phone use among drivers of passenger vehicles with no commercial markings observed between 8 AM and 6 PM at intersections controlled by a stop sign or stoplight. Estimates marked with an "H" or "L" are significantly high or low in their category.
[2] Change estimates marked with an "S" are statistically significant with 95% confidence.
Source: National Occupant Protection Use Survey, National Center for Statistics and Analysis, NHTSA

NCSA National Center for Statistics and Analysis, 400 Seventh St., S.W., Washington, DC 20590

5. NOPUS Survey Methodology

In this section we present details on the survey's methodology.

5.1 The Basic Survey Methodology

The NOPUS Sample

The results in this note were obtained from the Controlled Intersection Study of NOPUS. This survey uses a multi-stage probability sample of roadways to produce accurate estimates in a cost-efficient manner. The Controlled Intersection sample consists of intersections that are controlled by a stop sign or a stoplight, at which stopped and slowed traffic permit detailed observation. For a complete description of the NOPUS sample design, see (Glassbrenner, 2002).

Observation Protocols

Data collectors observed cell phone use and demographic characteristics of the drivers of passenger vehicles having no commercial or government markings that were stopped at a stop sign or stoplight during daylight hours between 8 AM and 6 PM. Observations were made both on surface streets and at the ends of expressway exit ramps (when there are controlled intersections). The 2002 survey was conducted between June 3, 2002 and June 22, 2002, and observed 38,000 vehicles.

Demographic classifications, as well as urbanization, were made according to the best determination of the data collectors. Rush hour is defined as the period 8:00-9:30 AM together with the period 3:30 – 6:00 PM on weekdays. See (Glassbrenner, 2002) for the States that comprise the four geographic regions.

An important piece of information regarding observation protocols is exactly what technologies and behavior were classified in NOPUS as "using a hand-held phone". This is described in Section 5.3.

Computation of Estimates

The field data is entered, edited, and missing values of certain variables (race, gender, and age) imputed. Estimates and sampling errors are computed incorporating the complex sample design. Although we plan to use direct estimation of the variances of differences in future surveys, the variances on the 2000-2002 differences in this note are based on the assumption that the 2000 and 2002 surveys are independent.

NCSA National Center for Statistics and Analysis, 400 Seventh St., S.W., Washington, DC 20590

5.2 Improvements in 2002

NOPUS observed 12,000 vehicles at 700 intersections in 2000, compared to 38,000 vehicles at 1,100 intersections in 2002. Note however that many standard errors did not decrease substantially in 2002, primarily because use rates increased. The coefficients of variation were lower on average in 2002 than in 2000. See (Glassbrenner, 2003) for more information on how the new observation sites were chosen.

In addition, we calculated a number of new estimates in this report, such as the total numbers of hours and miles driven by people talking on cell phones and the percent of wireless use that occurs behind the wheel, given in the Executive Summary. These calculations were derived in Section 6 and involve extrapolations using auxiliary data sources, such as the National Household Travel Survey.

5.3 What Does NOPUS Consider "Using a Hand-Held Phone"?

Data collectors were instructed to count a driver as "using a hand-held phone" if <u>the driver was holding what appeared to be a phone to his/her ear</u>.

Observers were not trained in the various types of wireless phones, and so the technologies they classified as "phones" likely reflected their own conceptions of what constitutes a "phone". In particular, individual observers might or might not have classified corded car phones, Personal Data Assistants, or (in the unlikely event they saw any) satellite phones, as "phones".

As noted in Section 2, this definition of "hand-held use" was chosen because it is simple (hence leading to fewer observational errors) and can be implemented in the context of observing traffic from the roadside. E.g. the definition purposely excluded activities such as manual dialing since this could be difficult to reliably observe from the roadside, as it can occur below the level of the windows and windshield from the observer's point of view.

However (as we also noted earlier) this can lead to somewhat odd distinctions as to what is considered "use". E.g. manually dialing is not considered to constitute use in NOPUS, but voice-activated dialing is.

We note that much of the activity that the NOPUS definition of "hand-held use" captures is conversational, i.e. speaking and listening to a person on the other end of the line. However we do not use the term "conversing on a hand-held phone" to describe the NOPUS estimates because this description is also not quite accurate, given that NOPUS counts voice-activated dialing and vocally navigating a customer service menu as "use".

NOPUS observers did not receive specific training in how to deal with situations in which they initially observed a driver not using a hand-held phone, and later (in the same stopped traffic) observed him/her to be holding a phone. In this situation, observers may have changed their initial assessment from "not using a hand-held phone" to "using a hand-held phone", or not.

NCSA National Center for Statistics and Analysis, 400 Seventh St., S.W., Washington, DC 20590

The 2002 survey does not observe whether hands-free devices are being used, although the 2004 survey has been expanded to include the observation of headsets into which the driver is speaking.

5.3 Design Aspects that Impact Data

Two limitations concerning the NOPUS data are worth particularly noting. First, note that the estimates in this report reflect **daytime** use only, whereas use patterns may well differ at night. Next note that NOPUS only observes vehicles that are **stopped** at a stop sign or stoplight. It is entirely conceivable that use rates are higher at such intersections, where e.g. drivers might decide to try to make a short phone call during a red light, than on general roadways. It is also conceivable that drivers might use hand-held phones more frequently on roads that have controlled intersections than on other roadways because they expect to be periodically stopped at lights or stop signs. Consequently, estimates from or based on NOPUS may overestimate use on general roadways, and may over- or underestimate use at general times of day. However because it is the only probability-based observational survey of driver cell phone use nationwide, NOPUS provides the best observed use rates available.

Estimates in this report on hands-free use that were derived in part from NOPUS inherit the same data limitations to some degree. For instance, our estimate that drivers used phones hands-free for 2% of their driving time in 2002 was derived based on both NOPUS, which reflects only daytime use, and telephone surveys, which may reflect what the respondents view as their "typical" behavior, and so may or may not reflect use at other times of day. The net effect of these survey limitations is, of course, unclear.

5.4 Assessing Significance

Because NOPUS uses a probability sample, one can calculate the error its estimates incur from observing use for a sample of roadways and times rather than for all roads and times. The actual quantity being estimated by a NOPUS estimate is within twice the standard error of the estimated value with 95% confidence. (Standard errors for sample-based estimates are provided in the tables in this report.) This computation can be used to determine whether differences, such as the difference in hand-held use in the West and Northeast, are statistically significant (with 95% confidence). See (Glassbrenner, 2002) for detailed examples of such calculations.

NCSA National Center for Statistics and Analysis, 400 Seventh St., S.W., Washington, DC 20590

6. Derived Estimates

In addition to estimates directly computed from NOPUS, this report contains estimates, which we refer to as "derived estimates", that are computed from NOPUS estimates in conjunction with estimates from other data sources. The derived estimates, and the auxiliary data sources used for each, are as follows:

Table 13: Derived Estimates and the Auxiliary Data Sources Used to Derive Them

Derived Estimate	Reference Year	Data Sources Used to Compute the Estimate (in Addition to NOPUS)
Estimates Concerning Driving Holding Phones		
Number of drivers on hand-held phones in a daytime snapshot	2000	• The 2001 National Household Travel Survey • The 1995 National Personal Travel Survey
	2002	• The 2001 National Household Travel Survey • The 1995 National Personal Travel Survey
Hours spent on hand-held phones while driving	2000	• The 2001 National Household Travel Survey • The 1995 National Personal Travel Survey
	2002	• The 2001 National Household Travel Survey • The 1995 National Personal Travel Survey
Miles driven while holding a phone	2000	• The 2001 National Household Travel Survey • The 1995 National Personal Travel Survey
	2002	• The 2001 National Household Travel Survey • The 1995 National Personal Travel Survey
Estimates Concerning Drivers Using Cell Phones Via Hands-Free Devices		
The percentage of drivers using phones hands-free in a daytime snapshot	2000	• (Boyle et al., 2001)
	2002	• (Royal, 2003) • (Stutts et al., 2003)
Number of drivers using phones via hands-free devices in a daytime snapshot	2000	• The 2001 National Household Travel Survey • The 1995 National Personal Travel Survey • (Boyle et al., 2001)
	2002	• The 2001 National Household Travel Survey • The 1995 National Personal Travel Survey • (Royal, 2003) • (Stutts et al., 2003)

NCSA National Center for Statistics and Analysis, 400 Seventh St., S.W., Washington, DC 20590

Derived Estimate	Reference Year	Data Sources Used to Compute the Estimate (in Addition to NOPUS)
Hours spent using phones via hands-free devices while driving	2000	• The 2001 National Household Travel Survey • The 1995 National Personal Travel Survey • (Boyle et al., 2001)
	2002	• The 2001 National Household Travel Survey • The 1995 National Personal Travel Survey • (Royal, 2003) • (Stutts et al., 2003)
Miles driven while using a phone hands-free	2000	• The 2001 National Household Travel Survey • The 1995 National Personal Travel Survey • (Boyle et al., 2001)
	2002	• The 2001 National Household Travel Survey • The 1995 National Personal Travel Survey • (Royal, 2003) • (Stutts et al., 2003)

In this section, we explain how each derived estimate in the above table was computed. This section is organized as follows: Section 6.1 describes the various auxiliary data sources in the above table. Section 6.2 computes several estimates of the extent of hand-held versus hands-free use, some of which will be used in the derived estimates. Section 6.3 computes estimates concerning trip characteristics. Finally Section 6.4 computes all derived estimates, using the results of Sections 6.1 – 6.3.

6.1 Data Sources

As seen in the previous table, all derived estimates in this report were computed from NOPUS and a small number of other data sources. We describe each auxiliary source below:

National Survey of Distracted and Drowsy Driving Attitudes and Behavior (Royal, 2003)
This is a nationwide telephone survey conducted by the Gallup Organization on behalf of NHTSA. In this survey, 4,000 drivers, ages 16 and older, were asked questions concerning driver distraction, including their use of cellular phones, and drowsy driving. This survey used a simple random sample obtained by Random Digit Dialing methods. Interviews were conducted between February and April of 2002.

Survey of Cell Phone Use in North Carolina (Stutts et al., 2003)
This is a statewide telephone survey conducted by researchers at the University of North Carolina. In this survey, 650 licensed North Carolina drivers, ages 18 and older were interviewed on issues of cell phone use while driving, including their own behavior. This survey used Random Digit Dialing methods to sample respondents, and interviews were conducted between mid-June and mid-July 2002.

NCSA National Center for Statistics and Analysis, 400 Seventh St., S.W., Washington, DC 20590

Motor Vehicle Occupant Safety Survey (Boyle et al., 2001)

This is a nationwide telephone survey conducted by the firm Schulman, Ronca & Bucuvalas on behalf of NHTSA. In this survey, 6,000 persons, ages 16 and older, were asked questions concerning issues of occupant protection and safety, including the use of cellular phones. This survey used a simple random sample obtained by Random Digit Dialing methods. Interviews were conducted between November 2000 and January 2001.

National Household Travel Survey (2001) and the National Personal Travel Survey (1995)

This survey, which changed names in 2001, is a nationwide survey conducted by Oak Ridge National Laboratory and sponsored by three Agencies within the Department of Transportation: the Bureau of Transportation Statistics, the Federal Highway Administration, and the NHTSA. In this survey, information is obtained from approximately 100,000 persons, ages 5 and older, through a combination of interviews and the keeping of travel logs. The survey collects a variety of information on trip characteristics, include trip length and duration, start and stop times, means of transportation, and trip purpose. Data from the 1995 survey were collected between May 1995 and July 1996, while the 2001 data was collected between April 2001 and May 2002.

6.2 The Extent of Hand-Held Versus Hands-Free Use

This section concerns the extent to which phones are used in some type of hands-free mode. Of course, some people do not have hands-free attachments to their cell phones (or other hands-free devices), or have them but do not use them. We will call these people "full time hand-held users". At the other extreme, some people have and use their hands-free attachments (or some other means of hands-free use) exclusively. We will call these, "full time hands-free users". In the sizable middle territory are people who have hands-free technology, and use it some, but not all of the time. These people might tend to use their phone hands-free when they drive, or their choice of whether to use their hands-free device might depend on other circumstances or preferences. We call these people "part-time hands-free users" (or just as fairly "part-time hand-held users").

Recall that NOPUS, at least for the data years in this report, observes the use of hand-held, but not hands-free phones. To extrapolate NOPUS's hand-held estimates to estimates concerning the use of cell phones via hands-free devices (such as the percent of drivers using phones hands-free, the number of drivers using phones hands-free at any given daylight moment, etc), we need information on the relative extent of hand-held versus hands-free use that occurs when people are driving.

A number of surveys, such as (Royal, 2003) and (Boyle et al., to appear), provide information on the distribution of hand-held versus hands-free use, but usually do so in a manner that is not most advantageous for NOPUS. These surveys commonly ascertain information on the means usually used. For instance, (Royal, 2003) measures the percentage of people who report usually holding a phone when they use a phone while driving, among people who have used a phone while driving, finding the following.

NCSA National Center for Statistics and Analysis, 400 Seventh St., S.W., Washington, DC 20590

Table 14: The Means of Cell Phone Usually Used While Driving in 2002

Means of Use	Percent of Drivers Who Report Usually Using This Means When They Use a Phone While Driving, Among Drivers Who Have Used a Cell Phone While Driving
Holding the phone	65%
Using the phone hands-free	35%
Total	100%

Source: (Royal, 2003)

We note that surveys indicate the increasing popularity of hands-free devices for using phones while driving. The Motor Vehicle Occupant Safety Survey estimated that in 2000, 73% of drivers who usually have a phone in their vehicle tend to hold the phone, 22% usually use a wireless device, and 5% either did not know or did not answer the question. (Boyle et al., 2001)

To extrapolate the NOPUS hand-held estimate to all wireless phones, a more helpful piece of information than that in the previous table would be the distribution of hand-held and hands-free use in a snapshot of drivers on phones. Or equivalently, we would like an estimated percentage of time that drivers spend holding phones, compared to those using hands-free devices.

Fortunately a study conducted at the University of North Carolina in 2002 fills in the gap. This study surveyed 650 drivers in North Carolina, ages 18 and older, between mid-June and mid-July in 2002 on various aspects of cell phone usage. It found the following information on the extent to which cell phone users use hands-free modes while driving.

Table 15: The Extent to Which Hands-Free Devices Were Used in North Carolina in 2002

Assertion Concerning Cell Phone Use While Driving	Percent of Drivers Who Fit the Assertion, Among Drivers Who Have Used a Cell Phone While Driving
I always hold the phone.	71.9%
I use hands-free devices 1-29% of the time.	3.5%
I use hands-free devices 30-59% of the time.	6.0%
I use hands-free devices 60-89% of the time.	5.2%
I use hands-free devices 90-99% of the time.	3.9%
I always use hands-free devices.	9.5%
Total	100.0%

Source: (Stutts et al., 2003)

See (Stutts et al., 2003) for the precise wording of survey questions. Note in particular that among the drivers in the Stutts study, 79.4% usually use hand-held phones when they use a phone while driving. (Here we have summed the entries 71.9%, 3.5%, and 2/3 of the entry 6.0% from the previous table.) However, this study was restricted to drivers in North Carolina. Nationwide, the Gallup survey in (Royal, 2003) estimates this percentage to be 65% (or 64.9% for more precision). That is, hands-free devices seem to be less popular in North Carolina than in the rest of the country. Calibrating the estimates in the previous table to be consistent with the

NCSA National Center for Statistics and Analysis, 400 Seventh St., S.W., Washington, DC 20590

Gallup national figure produces the following nationwide estimates for the extent of hands-free use.

Table 16: The Extent to Which Hands-Free Devices Were Used Nationwide in 2002

Assertion Concerning Cell Phone Use While Driving	Percent of Drivers Who Fit the Assertion, Among Drivers Who Have Used a Cell Phone While Driving
I always hold the phone.	58.8%
I use hands-free devices 1-29% of the time.	2.9%
I use hands-free devices 30-59% of the time.	4.9%
I use hands-free devices 60-89% of the time.	9.3%
I use hands-free devices 90-99% of the time.	7.1%
I always use hands-free devices.	17.1%
Total	100%

Data derived from the following sources: (Stutts et al., 2003), (Royal, 2003)

That is, we have multiplied the driver percentages corresponding to 59% hands-free time or less by 64.9/79.4, and have adjusted the remaining driver percentages by a constant multiple so that the estimates total to 100%.

We will assume that the people in this table who use hands-free devices for 1-29% of the time use the devices 15% of the time on average, and that the average use rates for the other categories in the table are the midpoints (45%, 75%, and 95%) of the corresponding time intervals. Taking a weighted average then gives that on average, part-time hands-free users use hands-free devices 67.6% of the time (that they use a phone while driving). Here, we have weighted the average percent of time in each part-time use assertion by the percent of drivers that fit the assertion. That is, we have computed the sum of the column denoted "D*T" in the following table for the rows in which hands-free units were used between 1% and 99% of the time (i.e. "part-time") and divided by the corresponding sum of column denoted "D".

NCSA National Center for Statistics and Analysis, 400 Seventh St., S.W., Washington, DC 20590

Table 17: Calculation of the Extent of Hands-Free Use Among Part-Time Hands-Free Users

Assertion Concerning Cell Phone Use While Driving	Percent of Drivers Who Fit the Assertion, Among Drivers Who Have Used a Cell Phone While Driving (D)	Avg % of Time on Hands-Free Phones in the Assertion (T)	Product (D*T)
I always hold the phone.	58.8%	0%	0.0%
I use hands-free devices 1-29% of the time.	2.9%	15%	0.4%
I use hands-free devices 30-59% of the time.	4.9%	45%	2.2%
I use hands-free devices 60-89% of the time.	9.3%	75%	7.0%
I use hands-free devices 90-99% of the time.	7.1%	95%	6.7%
I always use hands-free devices.	17.1%	100%	17.1%
Total	100%		
*The Percent of Time that Part-Time Hands-Free Users Use Hands-Free Devices (i.e. the sum of column D*T for those who use hands-free devices between 1% and 99% of the time divided by the sum of column D for the same people)*	**67.6%**		

Data derived from the following sources: (Stutts et al., 2003), (Royal, 2003)

Assuming that all groups in this table use phones for equal amounts of their driving time, one would expect a random snapshot of drivers using phones to consist of 58.8% full-time hand-held users, 16.3% part-time hands-free users using hands-free units, 7.8% part-time hands-free users using hand-held units, and 17.1% full time hands-free users.

Table 18: A Typical Snapshot of Drivers on Cell Phones Nationwide in 2002, by Type of User and Means of Use

Characteristic of Drivers in a Snapshot of Drivers Using Cell Phones	Percent of Drivers in the Snapshot that Meet the Characteristic
Drivers who never use hands-free devices (full-time hand-held users)	58.8%
Drivers who sometimes hold phones while driving, but are using a hands-free device in the snapshot (part-time hands-free users using hands-free units)	16.3%
Drivers who sometimes use hands-free devices while driving, but are holding a phone in the snapshot (part-time hands-free users using hand-held units)	7.8%
Drivers who always use hands-free devices to use a phone while driving (full-time hands-free users)	17.1%
Total	100%

Data derived from the following sources: (Stutts et al., 2003), (Royal, 2003)

The entries concerning full-time hand-held users and full-time hands-free users were copied directly from Table 15, while the entries for the part-time hands-free users were obtained by applying our newly derived 67.6% to the sum of entries concerning the part-time hands-free users from the same table.

Summing the entries by the type of user, we obtain the following distribution of the means used (hand-held or hands-free) to communicate on phones.

NCSA National Center for Statistics and Analysis, 400 Seventh St., S.W., Washington, DC 20590

Table 19: A Typical Snapshot of Drivers on Cell Phones Nationwide in 2002, by Phone Used

Characteristic of Drivers in a Snapshot of Drivers Using Cell Phones	Percent of Drivers in the Snapshot that Meet the Characteristic
Drivers on hand-held phones	66.6%
Drivers using phones hands-free	33.4%
Total	100%

Data derived from the following sources: (Stutts et al., 2003), (Royal, 2003)

That is, if one took a snapshot of all drivers on the road at some random time during 2002 and looked at the drivers on phones, one would expect to find 66.6% of them holding their phones, with the remaining 33.4% using phones hands-free.

Note the assumptions needed for this calculation. We assumed that the groups in Table 18 use phones for the same amount of time. E.g. we assume that the annual amount of time spent using a phone while driving is similar across groups. One group might drive more than another, or use phones for a smaller fraction of their driving time, but the total amounts spent on a phone while driving, in say, a year, must be similar. This assumption might be false. For instance, it might not be surprising if, e.g., people who own hands-free attachments make more phone calls than those who do not. However the estimates we have derived in this section are to our knowledge the best available.

Note that the calculations in this section strictly speaking concern all times of day, i.e. the "random time" of the snapshot is some random time of the day or night. However, since both (Royal, 2003) and (Stutts et al., 2003) rely on the recollection of survey respondents concerning their cell phone behavior, it is possible that respondents may report behavior based on what they think of as their "typical" phone calls. Perhaps they think of their daytime phone calls to home or business as their usual calls, or perhaps they think of calls to friends in the evening as typical. Consequently, our 66.6 : 33.4 distribution may reflect daytime or nighttime behavior to some greater extent. For our purposes, we will treat the 66.6 : 33.4 distribution as if it reflected daytime use. Because the Gallup and Stutts surveys did not specifically ask about daytime behavior, it was not possible to obtain a separate daytime distribution.

6.3 Trip Characteristics

In this section we calculate various statistics concerning characteristics of trips, such as the number of daytime trips in 2002, their average duration, and their average length in miles. These will be obtained using data from the National Household Travel Survey (formerly known as the National Personal Travel Survey), combined with data on vehicle registrations from R.L. Polk & Co. and the Federal Highway Administration.

The National Household Travel Survey (NHTS) changed names starting with the 2001 survey year. In previous data years (the next most recent is the 1995 survey), the survey was known as the National Personal Travel Survey (NPTS).

NCSA National Center for Statistics and Analysis, 400 Seventh St., S.W., Washington, DC 20590

We shall apply statistics from the NHTS and NPTS to NOPUS data, and so the reader should keep in mind that these surveys concern slightly different populations of vehicles. The NHTS and NPTS survey privately owned vehicles. For instance the NHTS estimates the number trips made each day in privately owned vehicles, but cannot say how many passenger vehicle trips occurred. In comparison, NOPUS observes passenger vehicles with no commercial or government markings. Consequently, a taxi driver who owns his/her cab will be counted in NHTS but not NOPUS, while an unmarked company car will be reflected in NOPUS but not NHTS. However, for the most part these surveys capture the same set of vehicles, and so we feel comfortable combining results from the surveys. To remind the reader, we will note the distinction in survey populations in the footers of tables and charts.

Using the online data analysis tools for the NHTS and NPTS available through the website for the Bureau of Transportation Statistics, we computed the following estimates. At the time of this report's publication, the data analysis tools could be found at http://nhts.ornl.gov/2001/index.shtml for the 2001 survey and at http://npts.ornl.gov/npts/1995/Doc/index.shtml for the 1995 survey.

Table 20: Trip Characteristics in 1995 and 2001, by Time of Day

Characteristic of Privately Owned Vehicles	1995 Estimates			2001 Estimates			Annualized Increase		
	Daytime*	Nighttime*	All Times of Day/Night	Daytime*	Nighttime*	All Times of Day/Night	Daytime*	Nighttime*	All Times of Day/Night
Average number of trips per day, in millions	419	211	629	444	200	644	1%	-1%	0%
Average duration of a trip, in minutes	15	17	16	18	20	19	3%	3%	3%
Average duration of trips per day, in millions of hours	106	61	167	134	66	200	4%	1%	3%
Average length of a trip, in miles	8	10	9	9	11	10	2%	2%	2%
Average length of trips per day, in millions of miles	3,525	2,179	5,703	4,102	2,214	6,317	3%	0%	2%
Average number of trips per vehicle per day	2	1	3	2	1	3	0%	0%	0%

[1] A daytime trip is defined to be a trip starting between 8 AM and 6 PM, with nighttime trips being those trips that start during other times.

Data derived from the following sources:
- The 2001 National Household Travel Survey, Dept. of Transportation
- The 1995 National Personal Travel Survey, Dept. of Transportation
- (Traffic Safety Facts 2001, 2002)
- (Traffic Safety Facts 1995, 1996).

Note the sometimes subtle distinction among the estimates in the above table. For instance, the average duration of a trip is the amount of time that elapses in a typical trip, while the average duration of trips per day is the total amount of time for all trips that occur in a typical day.

NCSA National Center for Statistics and Analysis, 400 Seventh St., S.W., Washington, DC 20590

We note that the NHTS and NPTS define a trip as traveling from one address to another using a single mode of transportation. A trip does not have to originate from home. Thus the average daytime trip of 9 miles means that on average when people get into a privately owned vehicle between the hours of 8 AM and 6 PM to reach a destination, they travel 9 miles to reach their destination.

In this table we have used data on vehicle registrations fom NHTSA's *Traffic Safety Facts* publications to compute the number of trips per vehicle. These publications estimate vehicle registrations using data from the Federal Highway Administration and R.L. Polk & Co.

Note in this table that we are defining a "daytime trip" to be a trip that starts between 8 AM and 6 PM. Note that since most trips are relatively short in duration (about 15-20 minutes), such trips corresponding roughly to trips that occur during daylight hours.

Of course the data years 2000 and 2002 are more relevant to our report, and we chose the 1995 and 2001 years of the NHTS (or NPTS) because they were closest to these years. Using the annual changes in the previous table, we derived corresponding estimates for 2000 and 2002 for the number of trips per day and the average length and duration of trips.

Table 21: Trip Characteristics in 2000 and 2002, by Time of Day

Characteristic of Privately Owned Vehicles	Daytime [1]		Nighttime [1]		Daily	
	2000	2002	2000	2002	2000	2002
Average number of trips per day, in millions	439	448	202	198	641	646
Average duration of a trip, in minutes	18	19	19	20	18	19
Average length of a trip, in miles	9	9	11	11	10	10
Average number of vehicles on the road at any given time, in millions	13	14	5	5	8	9
Average number of trip starts per hour, in millions	44	45	14	14	27	27

[1] A daytime trip is defined to be a trip starting between 8 AM and 6 PM, with nighttime trips being those trips that start during other times.
Data derived from the following sources:
- The 2001 National Household Travel Survey, Dept. of Transportation
- The 1995 National Personal Travel Survey, Dept. of Transportation
- (Traffic Safety Facts 2001, 2002)
- (Traffic Safety Facts 1995, 1996).

In this table we calculated the average number of vehicles on the road at any given, say, daylight, moment as follows. From this table, we know that the combined duration of all daytime vehicle

NCSA National Center for Statistics and Analysis, 400 Seventh St., S.W., Washington, DC 20590

trips in 2000 averaged 128 million hours per day, i.e. the number of daytime trips per day times their average duration. (Here we are using more digits than are presented in the table. Using the digits presented in the table, one obtains 132 million hours.) Since 10 daylight hours transpire each day, under our definition of daytime as 8 AM to 6 PM, there were on average 13.2 million vehicles on the road at any given time between 8 AM and 6 PM in 2000. The nighttime calculation is similar.

Also in this table, the number of trip starts per daylight hour was calculated simply by dividing the number of trips per day by the 10 hours that occur between 8 AM and 6 PM. The nighttime calculation is similar.

In particular, during the average daylight moment, about 14 million vehicles are on the road. This represents about 7% of licensed drivers and 6% of registered vehicles. (Traffic Safety Facts 2002, 2004) Of course, these averages reflect both peak driving times, such as rush hours, and non-peak times.

6.4 Applying these Estimates to NOPUS

Having obtained estimates on the extent of hands-free use and trip characteristics, we are now ready to apply these data to NOPUS. Recall that NOPUS estimates hand-held use to be 3% in 2000 and 4% in 2002 (during daytime, among passenger vehicles stopped at controlled intersections). In this section we will now produce several additional estimates that place these numbers, that some readers may have a hard time getting their hands around, into contexts that readers may find more meaningful.

6.4.1 Extrapolating NOPUS to Estimates of All Cell Phones

We start by extrapolating the NOPUS hand-held estimates to estimates for the hands-free use of phones and the use of cellular phones through either means. We derived in Section 6.2 that a snapshot of drivers on phones in 2002 would reveal (on average) 66.6% holding phones, and the remaining 33.4% using some type of hands-free mode. Applying this to NOPUS's estimate that 4% of drivers were holding phones then gives that 2% were using phones hands-free, for a total of 6% using phones through either means. That is, we have multiplied the NOPUS estimate of 4% by 33.4 / 66.6 to obtain the hands-free estimate.

Had Stutts et al. conducted a similar study in 2000 to their 2002 study (Stutts et al., 2003), we could have obtained a snapshot distribution of hand-held versus hands-free use for 2000. However no such study was done. The best source, which is somewhat limited for our purposes, to estimate such a distribution is the following information collected by NHTSA's 2000 Motor Vehicle Occupant Safety Survey (MVOSS).

NCSA National Center for Statistics and Analysis, 400 Seventh St., S.W., Washington, DC 20590

Table 22: The Means Usually Used While Driving to Use a Phone in 2000

Means of Use	Percent of Drivers Who Report Usually Using This Means When They Use a Phone While Driving, Among Drivers Who Have Used a Cell Phone While Driving[1]
Holding the phone	77%
Using hands-free devices	23%
Total	100%

[1]Unknowns have been distributed.
Source: (Boyle et al., 2001)

Note that the above table provides information on the phone usually used, not the hand-held / hands-free distribution we would expect to see in a snapshot. We would not feel comfortable adjusting this data using the Stutts study, as we did in Section 6.2 for the 2002 Gallup survey, because it is not unreasonable to suspect that cell phone use patterns may have changed in substantial ways in the period 2000 – 2002. Consequently, the NOPUS 2000 estimate of 3% hand-held use is adjusted using the estimates in the above table to yield 1% hands-free use, for a total of 4% use by cell phones in general. (This is the same way that the data was adjusted in (Utter, 2001).)

In summary we have extended NOPUS's national estimates of phones being held to the following estimates of phones used in a hands-free mode and phones used through either means.

Table 23: Driver Cell Phone Use Nationwide

Estimate	2000	2002	Annualized Increase
Percent of drivers using cellular phones[1]	4%	6%	22%
Drivers holding cell phones	3%	4%	15%
Drivers using cell phones hands-free	1%	2%	41%

[1] Drivers of passenger vehicles with no commercial markings observed between 8 AM and 6 PM at intersections controlled by a stop sign or stoplight.
Data derived from the following sources:
- The 2000 and 2002 National Occupant Protection Use Surveys, National Center for Statistics and Analysis, NHTSA
- (Royal, 2003)
- (Stutts et al., 2003)
- (Boyle et al., 2001)

We note that the NOPUS hand-held estimate was inflated by a larger amount in 2002 than in 2000 to produce the estimated hands-free use. The 2002 NOPUS hand-held data were inflated by $33.4/66.6 \times 100\%$, compared to $23/77 \times 100\%$ for 2000. The increase in the inflation factor reflects the increased popularity of hands-free devices during this time.

Recall that the data we used to adjust the NOPUS estimates (i.e. the 66.6 : 33.4 distribution of hand-held : hands-free use in 2002, and the 2000 MVOSS) reflect general times of day and night, whereas the NOPUS estimates reflect daytime use only. It is quite possible that hand-held :

NCSA National Center for Statistics and Analysis, 400 Seventh St., S.W., Washington, DC 20590

hands-free distributions differ substantially between day and night. Consequently our extrapolations in the previous table (i.e. the estimates of hands-free use and the use of phones through any means) should be viewed as rough estimates, although to our knowledge they are the best estimates available at the present time.

6.4.2 The Numbers of Vehicles with Drivers on Phones

Recall that in Section 6.3 we derived the estimated numbers of vehicles on the road at any given time (on average). Specifically we computed the average number of privately owned vehicles on the road during each moment between the hours of 8 AM and 6 PM. Applying our estimated percent of drivers on cell phones from the previous table translates these percentages into the average numbers of drivers using hand-held or hands-free cell phones use at any given time.

Table 24: Drivers on the Road at Any Given Time[1], By Phone Use

Estimate	2000	2002	Annualized Increase
Average number of drivers[1] on the road at any given time	12,847,815	13,919,148	4%
Drivers holding cellular phones[2]	385,434	566,788	21%
Drivers using cell phones hands-free[2]	116,158	283,965	56%
Drivers on cell phones[2]	501,593	850,753	30%
Drivers not on cell phones[2]	12,346,222	13,068,396	3%

[1] Drivers of privately owned vehicles on the road between 8 AM and 6 PM

[2] Based on observations of passenger vehicles with no commercial or government markings observed between 8 AM and 6 PM at intersections controlled by a stop sign or stop light.

Data derived from the following sources:
- The 2000 and 2002 National Occupant Protection Use Surveys, National Center for Statistics and Analysis, NHTSA
- The 2001 National Household Travel Survey, Dept. of Transportation
- The 1995 National Personal Travel Survey, Dept. of Transportation
- (Stutts et al., 2003)
- (Royal, 2003)
- (Boyle et al., 2001)

We note a key limitation of the above table: The number of drivers on the road is the number of privately owned vehicles, while the breakout of this number into those using cellular phones through various means is based on observations of passenger vehicles with no commercial or government markings at intersections controlled by a stop sign or stop light. Again we feel that the distinction between privately owned vehicles and vehicles with no commercial or government markings is small enough to warrant the calculation, and we caution the reader, as we did in the Executive Summary, that cell phone use might be higher or lower at controlled intersections than at general roadway sites.

NCSA   National Center for Statistics and Analysis, 400 Seventh St., S.W., Washington, DC 20590

6.4.3 Relating Phone Use to Trip Length and Duration

We can similarly relate cell phone use to other trip characteristics. In Section 6.3 we derived a number of statistics concerning characteristics of trips. For instance, we calculated the average duration of a daylight trip (19 minutes in 2002) and the average distance traveled during a daylight trip (9 miles in 2002). Applying our estimated percent of drivers on cell phones from the previous table expresses the amount of driver cellular phone use taking place in terms of these trip characteristics.

Table 25: Driver Cell Phone Use, As It Relates to the Distance and Duration of Trips

Estimate	2000	2002	Annualized Increase
Average duration of a trip[1], in minutes	18	19	3%
Minutes with driver holding a phone[2]	0.5	0.8	26%
Minutes with driver using a phone hands-free[2]	0.2	0.4	41%
Minutes with driver on a phone[2]	0.7	1.1	25%
Minutes with driver not on phone[2]	16	16	0%
Average length of a trip[1], in miles	9	9	0%
Miles traversed with driver holding a phone[2]	0.3	0.4	15%
Miles with driver using a phone hands-free[2]	0.1	0.2	41%
Miles with driver on a phone[2]	0.4	0.6	22%
Miles with driver not on phone[2]	8	8	0%
Average time spent on trips[1] each day, in millions of vehicle-hours	128	139	4%
Time with driver holding a phone[2]	4	6	22%
Time with driver using a phone hands-free[2]	1	3	73%
Time with driver on a phone[2]	5	9	34%
Time with driver not on phone[2]	118	122	2%

NCSA National Center for Statistics and Analysis, 400 Seventh St., S.W., Washington, DC 20590

Estimate	2000	2002	Annualized Increase
Average distance traversed in trips[1] each day, in millions of vehicle-miles	3,984	4,224	3%
Distance traversed with driver holding a phone[2]	120	172	20%
Distance traversed with driver using a phone hands-free[2]	36	86	55%
Distance traversed with driver on a phone[2]	156	258	29%
Distance traversed with driver not on phone[2]	3,673	3,708	0%

[1] Trips in privately owned vehicles that commence between 8 AM and 6 PM.

[2] Based on observations of passenger vehicles with no commercial or government markings observed between 8 AM and 6 PM at intersections controlled by a stop sign or stop light.

Data derived from the following sources:

- The 2000 and 2002 National Occupant Protection Use Surveys, National Center for Statistics and Analysis, NHTSA
- The 2001 National Household Travel Survey
- The 1995 National Personal Travel Survey

The above table in a sense expresses the NOPUS 4% use rate for hand-held wireless phones in the more concrete terms of distance and duration. In this table we see that during the average daylight trip, which is 19 minutes in duration and traverses 9 miles, the driver is on a hand-held phone for 0.8 of those minutes and 0.4 of those miles, on average. Note that these averages include drivers on the phone for some or all of their trip and those who make no calls on their trips. Over the course of 2002, drivers used hand-held phones for a total of 6 million daylight hours, during which they traversed 172 million miles.

Note that this table indicates that drivers holding phones traveled at about 30 mph in 2002, with a similar speed in 2000. However these speed estimates are somewhat contrived and should not be considered accurate. The estimated duration and distance of trips using hand-held phones in this table were obtained by applying the NOPUS 4% usage rate to the duration and distance of trips in general. Consequently, the estimated speed while holding a phone inherited the average speed of trips in general, which was about 30 mph, in this calculation. However it would be reasonable to think that people use their phones more when they are traveling at slower speeds, especially when they hold a phone. Recall in fact that NOPUS observes *stopped* vehicles, not vehicles in motion. (See Chapter 5 for more information on NOPUS's observation protocols.) Applying NOPUS's usages rates from stopped vehicles to vehicles in motion, as we have done in this table, may overstate or understate cell usage in vehicles in motion.

As we have mentioned with previous tables, the estimates in the above table have a limitation other than the possibility that cell phone could well be higher at controlled intersections than at general roadway sites. The trip characteristics in this table are based on privately owned vehicles, while the cell phone use information is based on observations of passenger vehicles with no commercial or government markings at intersections controlled by a stop sign or stop light.

NCSA   National Center for Statistics and Analysis, 400 Seventh St., S.W., Washington, DC 20590

7. References

Boyle, J., Vanderwolf, P., *2000 Motor Vehicle Occupant Safety Survey, Volume 4, Crash Injury and Emergency Medical Services Report*, NHTSA Technical Report, DOT HS 809 459, November 2001

Boyle, J., Vanderwolf, P., *2003 Motor Vehicle Occupant Safety Survey, Volume 4, Crash Injury and Emergency Medical Services Report*, NHTSA Technical Report, to appear.

Cingular Wireless, *Guys Still Gab More on Wireless*, report posted at http://www.cingular.com/about/latest_news/02_06_24

CNN.com, *Cell Use to Nearly Double by 2006*, September 17, 2002

Glassbrenner, D., *Safety Belt and Helmet Use in 2002 – Overall Results*, NHTSA Technical Report, DOT HS 809 500, September 2002

Glassbrenner, D., *Safety Belt Use in 2002 – Demographic Characteristics*, NHTSA Research Note, DOT HS 809 557, March 2003

Governor's Highway Safety Association, http://www.ghsa.org/.

Highway Statistics 2001, Federal Highway Administration, U.S. Department of Transportation, November 2002

National Household Travel Survey, website analysis tool from http://nhts.ornl.gov/

Roche, R., Jobanputra, P., Rodriguez, L., *CTIA's Wireless Industry Indices*, Cellular Telecommunications & Internet Association, April 2003

Royal, D., *National Survey of Distracted and Drowsy Driving Attitudes and Behaviors: 2002, Volume I – Findings Report*, NHTSA Research Note, DOT HS 809 566, April 2003

Stutts, J., Hunter, W., Huang, H., *Cell Phone Use While Driving: Results of a Statewide Survey*, Transportation Research Board, Annual Meeting CD-ROM, 2003

Traffic Safety Facts 1995: A Compilation of Motor Vehicle Crash Data from the Fatality Analysis Reporting System and the General Estimates System, National Center for Statistics and Analysis, NHTSA, DOT HS 808 471, 1996

Traffic Safety Facts 2001: A Compilation of Motor Vehicle Crash Data from the Fatality Analysis Reporting System and the General Estimates System, National Center for Statistics and Analysis, NHTSA, DOT HS 809 484, December 2002

Traffic Safety Facts 2002: A Compilation of Motor Vehicle Crash Data from the Fatality Analysis Reporting System and the General Estimates System, National Center for Statistics and Analysis, NHTSA, DOT HS 809 620, January 2004

NCSA National Center for Statistics and Analysis, 400 Seventh St., S.W., Washington, DC 20590

Utter, D., *Passenger Vehicle Driver Cell Phone Use: Results from the Fall 2000 National Occupant Protection Use Survey*, NHTSA Research Note, DOT HS 809 293, July 2001

NCSA National Center for Statistics and Analysis, 400 Seventh St., S.W., Washington, DC 20590

DOT HS 809 580
December 2004

U.S. Department
of Transportation

**National Highway
Traffic Safety
Administration**

People Saving People
www.nhtsa.dot.gov

www.ingramcontent.com/pod-product-compliance
Lightning Source LLC
Chambersburg PA
CBHW081357170526

45166CB00010B/3117